I0410073

SEEDS
in the Garbage

Naira R. Matevosyan, M.D, Ph.D

ISBN: 978-1466490758

To my son and all, who never cease to admire
and study the mysteries of Mother-Nature.

1

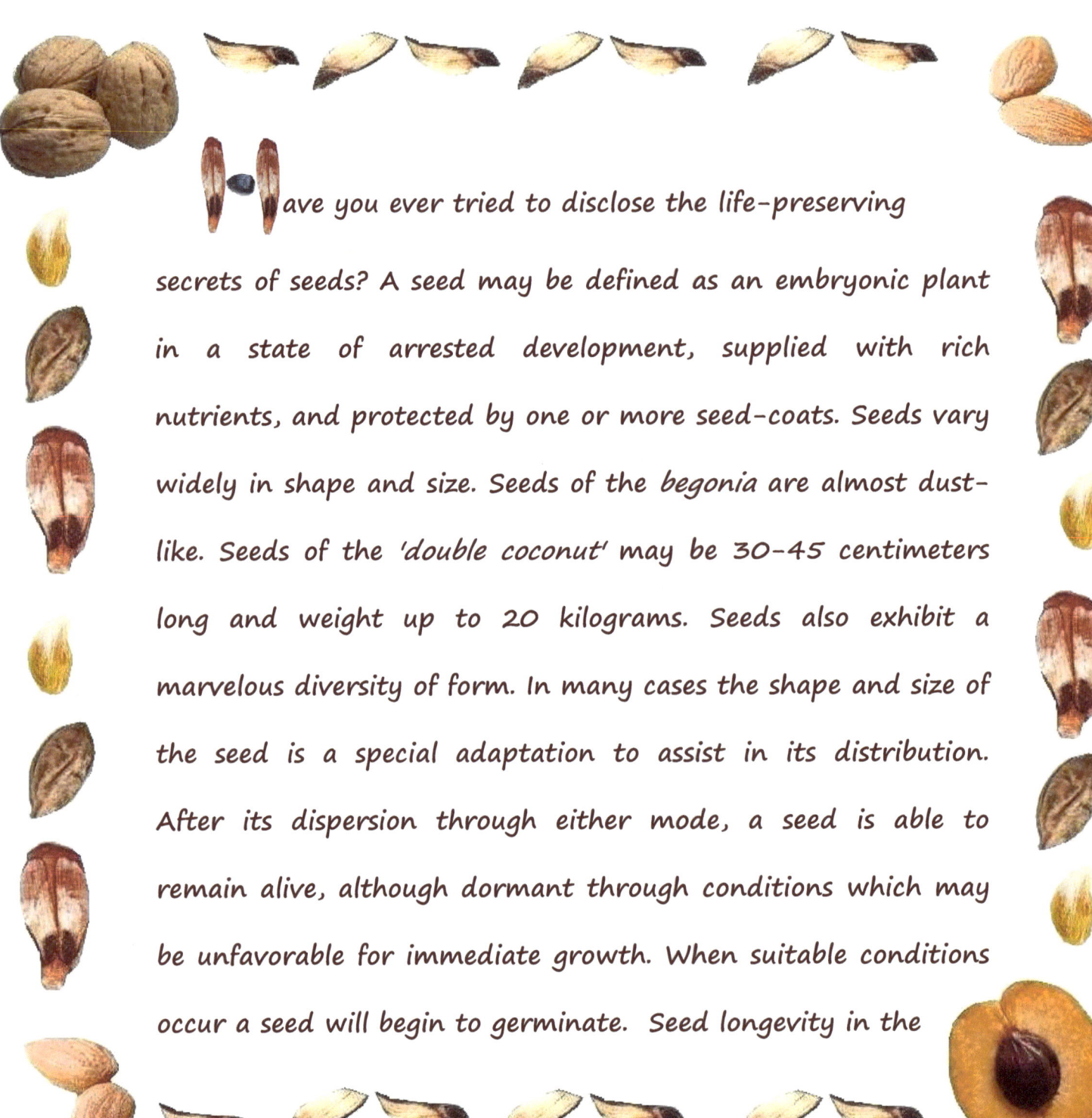

Have you ever tried to disclose the life-preserving secrets of seeds? A seed may be defined as an embryonic plant in a state of arrested development, supplied with rich nutrients, and protected by one or more seed-coats. Seeds vary widely in shape and size. Seeds of the *begonia* are almost dust-like. Seeds of the *'double coconut'* may be 30-45 centimeters long and weight up to 20 kilograms. Seeds also exhibit a marvelous diversity of form. In many cases the shape and size of the seed is a special adaptation to assist in its distribution. After its dispersion through either mode, a seed is able to remain alive, although dormant through conditions which may be unfavorable for immediate growth. When suitable conditions occur a seed will begin to germinate. Seed longevity in the soil depends on the interaction of many factors including the intrinsic dormancy of the seed population, the environmental conditions (light, temperature, moisture) and biological processes (predation, allelopathy).

ermination is indeed a fascinating process. Seeing a tiny seedling emerge from a dry, wrinkled seed and watching its growth and transformation is observing the mystery of life unfolding. Before the germination, seeds are dispersed to suitable conditions. Let's discuss the modes of dispersal, including the human factor.

lants, being stationary, require a mobile mode for seed dispersal. Modes include self-projectile mechanisms, gravity, ballistics, wind, water, plant-plant interaction, animals, and humans [Bullock et al, 2002]. Gravity is a simple mean of achieving seed dispersal. The effect of gravity on heavier fruits causes them to fall from the plant when ripe. Fruits exhibiting this type of dispersal include apples, coconuts and passionfruit and those with harder shells often roll away from the plant to gain further distance. Gravity dispersal also allows for later transmission by water or animal. Heavy seeds – for example, hazelnuts, acorns and chestnuts–are normally quite featureless, lacking structures such as hooks or wings. For this reason they usually just stay on the ground where they fall, which is not usually conducive to germination. For each of these seeds to become a tree, it needs to go to a lighted place where it can easily develop. Interestingly enough, jays, crows, woodpeckers, and most importantly, squirrels like eating these fruits and are the essential in the survival of oak and chestnut forests. These little creatures that collect the maturing seeds store them in various places, then forget where they have left those seeds. This 'amnesia' is the meaningful part of the ecosystem that creates symbiotic relationship between these two living things. At the storage, the nuts then germinate and grow into trees. Reliance on <u>wind</u> dispersal is common among many weedy or ruderal species. The 'wings' (samara) of the *maple tree* and the 'parachutes' (feathery pappus) of dandelions, milkweeds, and poplar are adapted for wind transportation and can be dispersed long distances. Species of many aquatic (water lily, lotus, cattail, mangrove tree) and terrestrial (palm tree) plants use seed dispersal through <u>water</u> (*hydrochory*). Their seeds are buoyant, waterproof and can withstand water transportation for extremely long distances. *Lys de mer, a* plant that grows on Mediterranean shores, has slightly angular seeds. When the outer case of the seeds matures, it takes on a mossy appearance and helps the seeds to be dispersed by floating on water. The seeds of the field pennycress are carried by rainwater. They are marked with little scratches like fingerprints, which serve to increase the surface tension by which the seeds are easily distributed. <u>Ballistic</u> or self-dispersal (*autochory* or *ballistochory*) is the physical and often explosive discharge of seeds from the fruit. The seeds are typically ejected from the fruit by elastic contraction of the fruit tissues and often the fruits are shaped such that seeds are flung away from the parent plant (violet, jewelweed, witch hazel). The fruits of *Hura crepitans* are segmented like a tangerine and sound like a gun shot when they explode. The flattened inflorescences of *Dorstenia* squeeze out seeds and shoot them several inches. The exploding seeds pods of *Bauhinia purpurea* (the orchid tree) can throw seeds nearly 50 feet. Evening primrose is another plant with ballistic knowledge. Its seeds are stored in capsules which are sealed when dry. When these capsules get wet, they immediately open in the shape of a goblet. In this position, raindrops are enough to distribute the seeds. While the ballistic dispersal does not often achieve the same distance as animal-dispersed seeds, many ballistic dispersed seeds also have a form of secondary dispersal. Some common examples of species employing ballistic dispersal include the aptly named *Touch-me-nots* whose fruits explosively dehisce and squirting cucumbers that discharge their seeds in a mucilaginous stream of liquid. Geranium species have a fruit capsule of five cells which end in cups that hold the seed. The cups are joined to a long beak-like column, and when the fruit is ripe, it springs open casting away the seeds like a catapult. Just for the record, the world champion ballistics title belongs to an African tree in the Legume family, *Tetraberlinia moreliana*, which throws its seeds almost 200 feet. This 'natural rocket system' is also common for the Mediterranean squirting cucumber that generates its own forces to distribute seeds. As it ripens, it fills with a slimy juice, which gradually creates pressure until the cucumber bursts off its stalk. Behind it comes a trail of slime like the trail behind a space rocket. By this means, the cucumber's seeds are dispersed on the ground together with the slime *[Ibid]*. The broom seed is an example of dispersion through the <u>plant-plant interaction</u>. The broom is another plant that reproduces by opening its seedpod of its own accord, but in a completely different way from that of the Mediterranean squirting cucumber. The broom's pods burst as a result of evaporation rather than as a result of an increase of liquid in the plant. As the heat rises, the side of the pod facing the sun dries out faster than that in the shade, which creates a tension in the pod. Finally it splits suddenly into two halves and its tiny black seeds are dispersed in all directions *[Ibid]*. The incredible twisting mechanism of Erodium is another example. The fruits of Erodium come together on their syles at one central point. The seeds are located inside the fruits. At maturity, the stamen attached to the seed starts to curl, extending towards the ground. This is when the plant's amazing mechanism comes into play, letting its seeds screw themselves into the soil. Plants that depend on <u>animals</u> for dispersal have seeds that are adapted to travel on the outside or the inside of the animal (mostly mammal), a process known as *epizoochory*.

Seeds with burrs, spines, bristles, or hooks can attach to an animal's fur (beggar-ticks, sandbur, burdock, dock). A typical example of an epizoochorous plant is *Trifolium angustifolium*, a species of the Old World clover which adheres to animal fur by means of stiff hairs covering the seed. Seed dispersal via ingestion and digestive waste by vertebrate animals (mostly birds and mammals), or *endozoochory*, is the dispersal mechanism for most tree species. For fleshy covered seeds transported internally (contaminated seeds), plants provide an attractive fruit pulp reward in return for the ride (apple, cherry, juniper). For example, seeds of tomato or blackberry have impervious coats which enable them to survive digestion in animals' stomachs. In the tropics, large animal seed dispersers (tapirs, chimpanzees, hornbills) may distribute large seeds with few other seed dispersal agents. Seed dispersal by ants (myrmecochory) is a dispersal mechanism of many shrubs of the southern hemisphere (Apalachian salamanders, tropical white eyes) or understorey herbs of the northern hemisphere (bloodroot, trillium). Seeds of myrmecochorous plants have a lipid-rich attachment called the *elaiosome*, which attracts ants [Beattie et al, 1981]. Ants carry such seeds into their colonies, feed the elaiosome to their larvae, pry off the tasty bit, and discard the otherwise intact seed in an underground chamber. This leaves the seed safely underground in an ant-midden, ready to germinate – a great way to dodge seed-eating critters and avoid competition from its parent plant and siblings [Giladi, 2006]. Evidence from genetic studies shows that limited seed dispersal in myrmecochory can lead to a strong genetic structure within populations even at spatial scales as small as a few meters. The failure of myrmecochores to maintain gene flow across barriers may lead to reproductive isolation of sub-populations, which may facilitate speciation [Moyle et al, 2009]. Thus, myrmecochorous plants make lots of new species not because their unique characteristics give them some adaptive advantage (although, to be sure, there are advantages to ant dispersal), but because ants do a lousy job moving seeds between populations, leaving them free to follow their own evolutionary trajectories. Lengyel et al [2009] argue that myrmecochory is a *key innovation*, a trait that helps a group of organisms spread and diversify in the process evolutionary biologists call *adaptive radiation*. Based on their results, ant dispersal is strongly associated with evolutionary diversification. But the speciation that myrmecochory promotes is an accident, a side effect. We often think of key innovations promoting speciation by adaptive means, by allowing one group of species to outcompete others. Clearly, however, a key innovation can also be a trait that makes the accident of speciation a little more likely. Dispersal by <u>humans</u> used to be seen as a form of dispersal by animals. Humans may disperse seeds by various means and some surprisingly high distances have been repeatedly measured (cherry, bean, wheat). It is often suggested that dispersal of seeds by humans may lead to non-native species invasions. A study led by *Wichmann et al [2008]* provides the first quantitative evidence that humans may be more important than natural agents, such as wind power, for the dispersal of plants across the landscape. The research was carried out by scientists from the Centre for Ecology & Hydrology together with colleagues from Utrecht University in the Netherlands and the University of Marburg

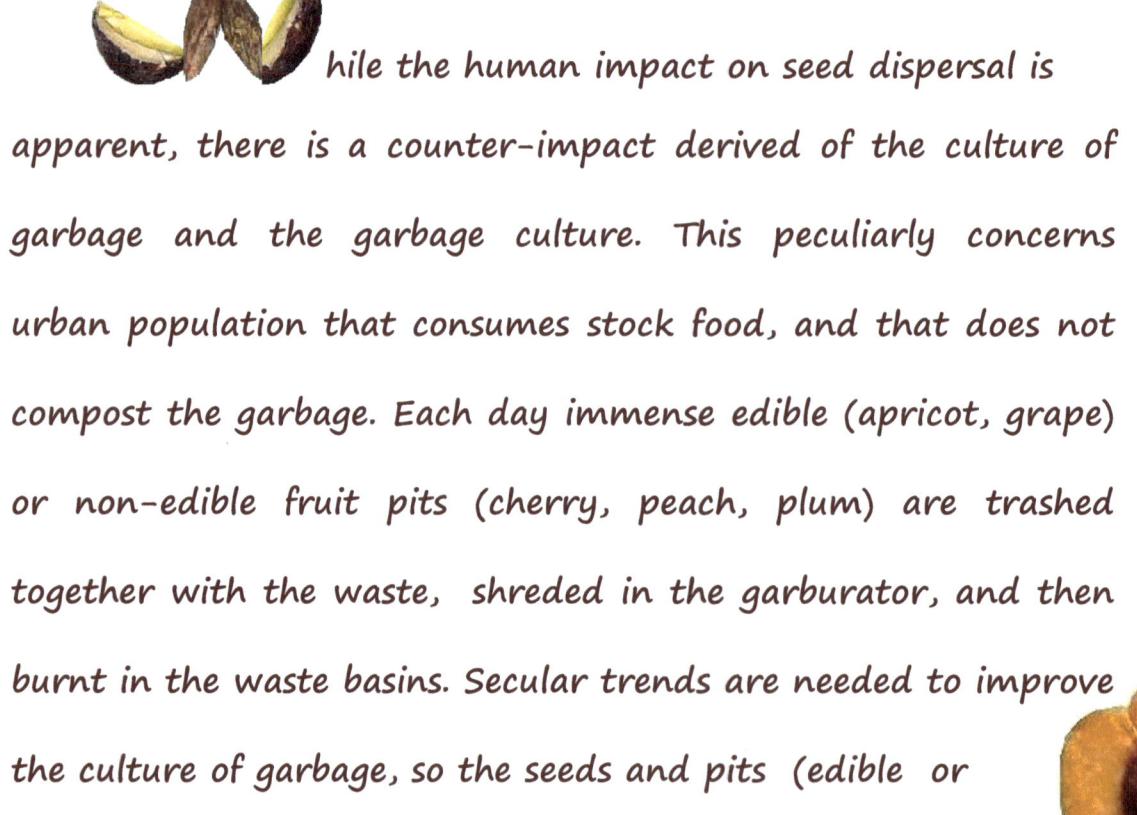

in Germany. Findings suggest that while wind dispersal takes seeds a few metres (less than 250m), adhesion to shoes or hikers` boots of walking humans disperses seeds over ten kilometres and thus, may colonize new sites. This is called a livestock (walking) dispersal *[Menalled, 2008]*.

hile the human impact on seed dispersal is apparent, there is a counter-impact derived of the culture of garbage and the garbage culture. This peculiarly concerns urban population that consumes stock food, and that does not compost the garbage. Each day immense edible (apricot, grape) or non-edible fruit pits (cherry, peach, plum) are trashed together with the waste, shredded in the garburator, and then burnt in the waste basins. Secular trends are needed to improve the culture of garbage, so the seeds and pits (edible or non-edible) are separated from the waste and returned to the soil in their either stage and condition (dormancy, rest period, pregerminative period). The disease healing and preventive potential of the fruit pits is enormous. If the seed returned to the soil does not germinate due to injuries, or because of the dormancy period, it at least nourishes the soil and humus with unreplacable vitamins, phosphates, anticancer, antiallergic, antimicrobial compounds transferred then to plants, crop, and harvest that we later consume, or trasferred to the milk or meat of the cattle, goats, or sheep that consumes the harvest. This circulation of bio-benefits is called ecosystem that functions in an absolute integrity and efficiency unless disturbed by anthropogenic selection or stress. Let`s face: we never discard coffee beans with garbage, but we discard kernels/pits of fruits that we consume each day.

This is a selective deterioration of Mother Earth`s potential, our life-time supplier of harvest, crop, and positive energy, and the releaser of our negative energy. There is no waste in nature: Mother-nature functions with the reason and accountability. In nature every single raindrop, spider, cocoon, leaf, spore or seed, counts. In that harmonious web there is no small detail and there is no role too big or too small.

Human idiopathic (unclear causality) infertility and sterility rates are periodically reported as 10% [Plaseski et al, 2011; Vog et al, 1996]. Analogically, from the perspective of the natural selection we may speculate that 10% of the fruit seeds will not sprout (for being infecund) even in ideal conditions for germination. This 10% of seeds is for nourishing the soil. When 'screening' the seed closely and studying its morphology, ask yourself a simple question whether or not, a viable, fertile seed and optimal conditions for germination (moisture, warmth, light) are enough to transform a tiny grain into a mature plant, bush, or tree? There is something else, something mysterious that breathes life and vigor into the transformation, into a new development. Name it the God`s blessing, or Natural reward. In any event, we have to appreciate that each seed contains a dream. Do not discard it to the waste. When your cat gives birth to kittens, you put efforts to find a loving heart and to rehome them. The same is with the fruting plant seeds: they need to be returned to the earth seedbank, rather then processed through the garburator. The earth is the permanent possessor of all things contained therein and which grow therefrom. Further, the more edible crop or fruit seeds are back to the soil bank, the less efforts

are needed in crop-weed competition.

 hen looking at a seed one looks at the 'seed coat' or testa. Seed coats perform much the same function as the coats we wear for protecting ourselves from the foul

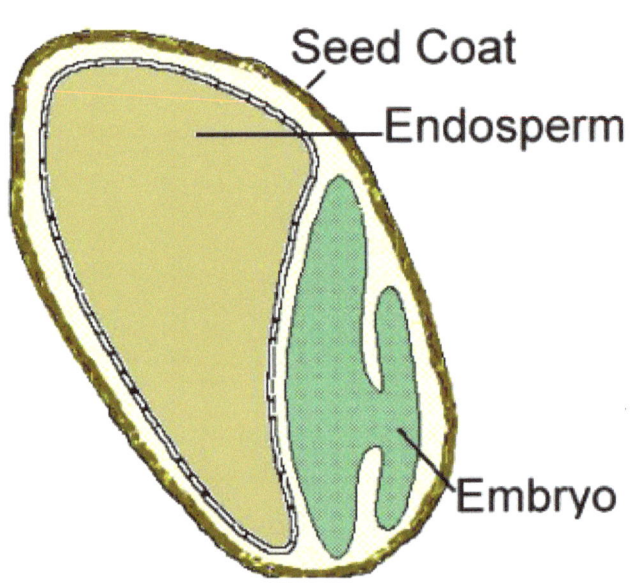

Seed Coat
Endosperm
Embryo

weather. Seed-coats provide protection against the entry of parasites, mechanical injuries, and unfavorably high or low temperatures. The coat of the mature seed can be a paper-thin

layer (peanut) or something more substantial (thick and hard in honey locust, coconut, or Kentucky coffee). In addition to the three basic parts, some seeds have an appendage on the coat

ARIL

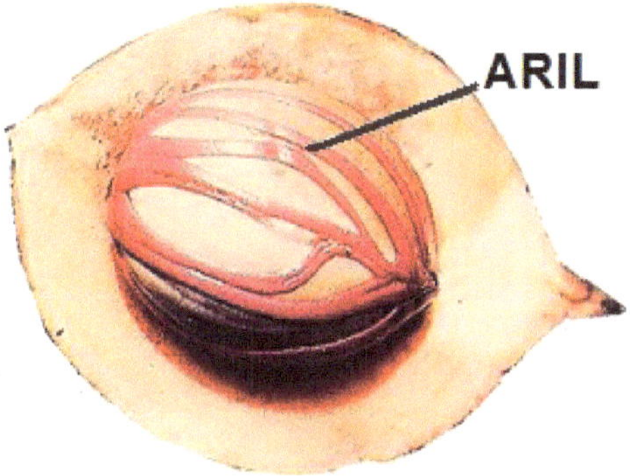

ARIL

Elaiosome

SEED   HAIR

such an aril (yew, nutmeg), an *elaiosome* (Corydalis), or hairs (cotton). These accessories and adaptations are designed to attract or facilitate seed dispersion. For example, the nutmeg aril attracts toucans who then disperse the seeds through regurgitation. The elaiosome is a dispersal unit as in the majority of myrmecochores *[Gorb et al, 2003]*. It has rich fatty inclusions that protect the seed from injuries while transported by the ants. The seed hairs ease the seed's passage into the soil through wind.

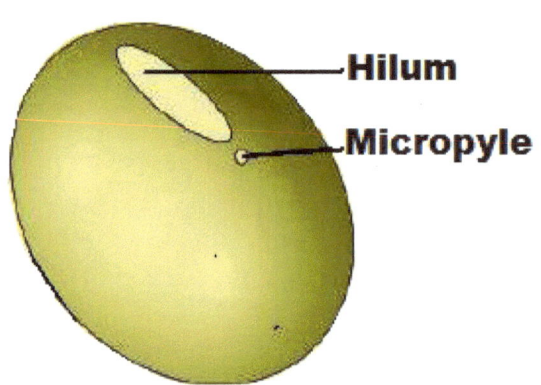

There may also be a scar on the seed coat, the *hilum*; it is where the seed was attached to the ovary wall by the funiculus. There is a tiny hole in the testa called the *micropyle*. When the seed is ready to germinate, water is taken in through the micropyle. Also, the first root, the radicle, will grow out of the seed through the micropyle.

Inside the seed-coat is the embryo, an immature plant with all the parts of the adult plant. A close look shows leaves and a root, the beginnings of a plant. The seed's embryo leaves are called the *cotyledons*. The seed is filled with *endosperm*, the

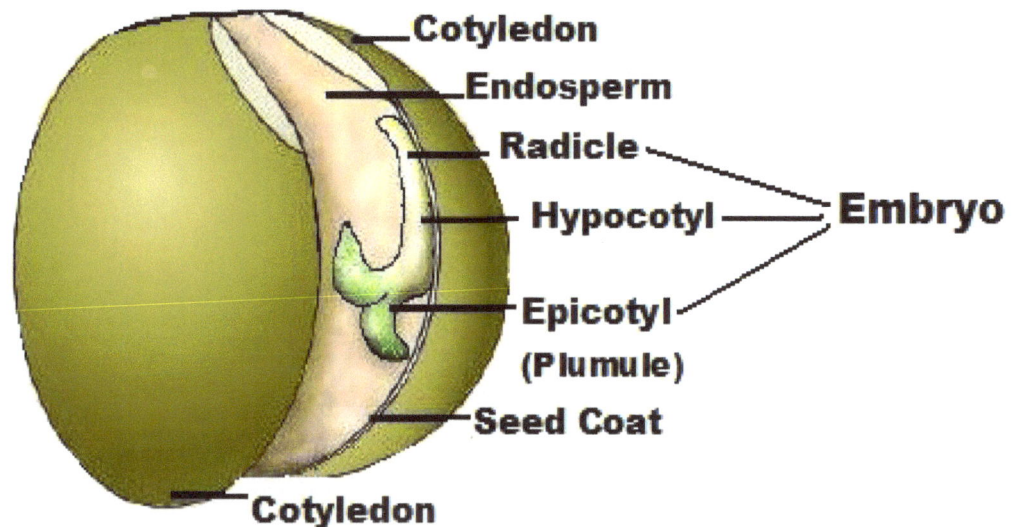

food that nourishes the embryo during its early stages of development. The *radicle* is the embryonic root. The *plumule* is the embryonic shoot. The embryonic stem above the point of attachment of the *cotyledon(s)* is the *epicotyl*. The embryonic stem below the point of the attachment is the *hypocotyl*. Within the seed, there usually is a store of nutrients for the seedling that will grow from the embryo. The form of the stored nutrition varies depending on the plant type. In *angiosperms* the stored food begins as a tissue called the endosperm, which is derived from the parent plant via double fertilization. The usually triploid endosperm is rich in oil or starch and protein. In *gymnosperms*, such as conifers, the food storage tissue is part of the female gametophyte, a haploid tissue. In some species, the embryo is embedded in the endosperm or *female gametophyte*, which the seedling will use upon germination. In others, the endosperm is absorbed by the embryo as the latter grows within the developing seed, and the cotyledons of the embryo become filled with this stored food. At maturity, seeds of these species have no endosperm and are termed *exalbuminous seeds*. Some exalbuminous seeds are bean, pea, oak, walnut, squash, sunflower, and radish. Seeds with an endosperm at maturity are termed *albuminous seeds*. Most monocots (grasses and palms) and many dicots (brazil nut and castor bean) have albuminous seeds. All gymnosperm seeds are albuminous.

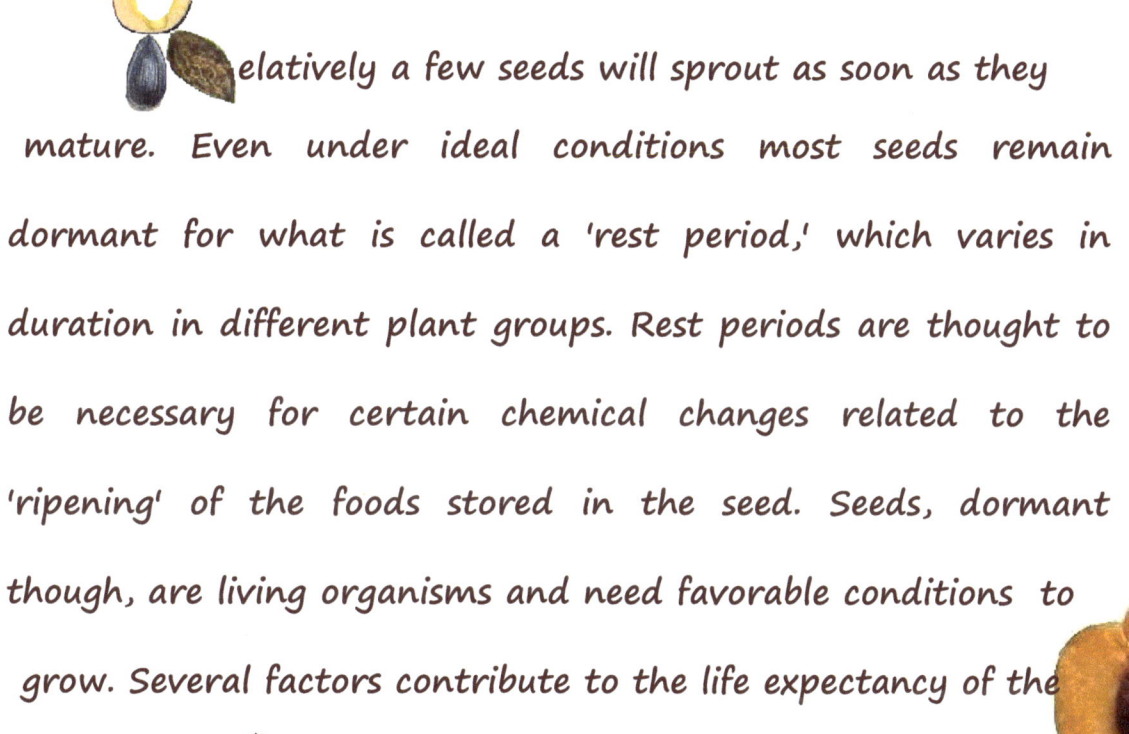

elatively a few seeds will sprout as soon as they mature. Even under ideal conditions most seeds remain dormant for what is called a 'rest period,' which varies in duration in different plant groups. Rest periods are thought to be necessary for certain chemical changes related to the 'ripening' of the foods stored in the seed. Seeds, dormant though, are living organisms and need favorable conditions to grow. Several factors contribute to the life expectancy of the seeds. The length of time a seed will remain viable (able to sprout strongly and produce sturdy seedlings) varies widely with the species and the care taken in harvesting and storing the seeds. Some seeds (chervil), rarely retain their germinating power longer than one year. Other seeds (celery, cabbage, cucumber, and various other vegetables and flowers) may sprout well when ten years old or more. Seeds not fully mature when gathered or stored before they are fully dry, or are kept at too high a temperature, may sprout poorly or not at all.

he first sign of germination is the absorption of water. This activates an enzyme, respiration increases and plant cells are duplicated. Soon the embryo becomes too large, the seed coat bursts open and the growing plant emerges. The tip of the root is the first thing to emerge and it's first for good reason. It will anchor the seed in place, and allow the embryo to absorb water and nutrients from the surrounding soil.

hat happens to the seeds that do not sprout? Some seeds are water-permeable, the others not (water-impermeable). Water-permeable seeds have morphological, physiological, or morphophysiological dormancy, with

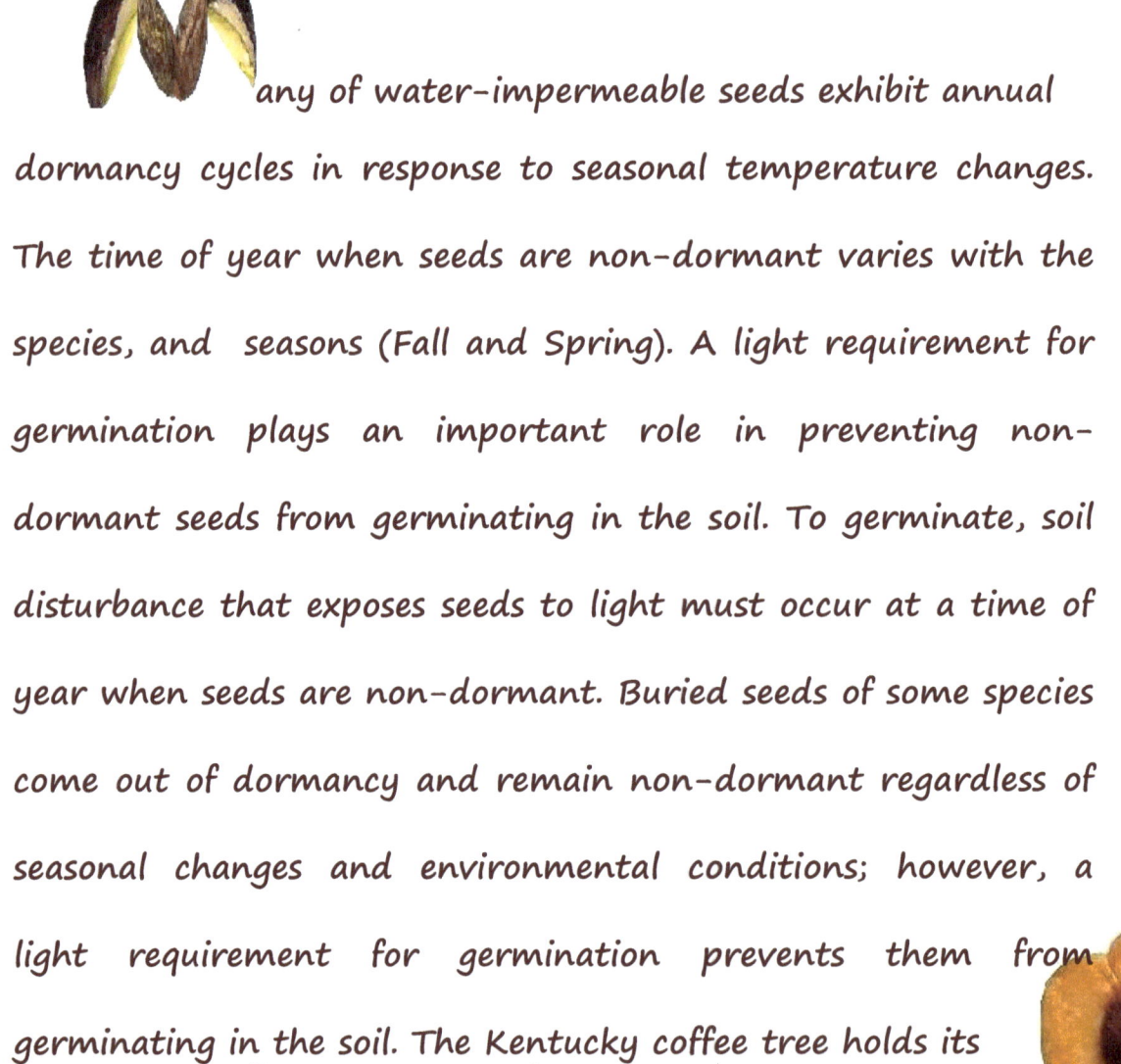

physiological dormancy being the one most commonly found in buried weed seeds *[Baskin et al, 2006]*.

Many of water-impermeable seeds exhibit annual dormancy cycles in response to seasonal temperature changes. The time of year when seeds are non-dormant varies with the species, and seasons (Fall and Spring). A light requirement for germination plays an important role in preventing non-dormant seeds from germinating in the soil. To germinate, soil disturbance that exposes seeds to light must occur at a time of year when seeds are non-dormant. Buried seeds of some species come out of dormancy and remain non-dormant regardless of seasonal changes and environmental conditions; however, a light requirement for germination prevents them from germinating in the soil. The Kentucky coffee tree holds its seed pods in the top of the tree all winter. The inside of the pod is fleshy (lots of water). The pods are very dark in color. If you put the fickle winter and sunshine and darkness into this

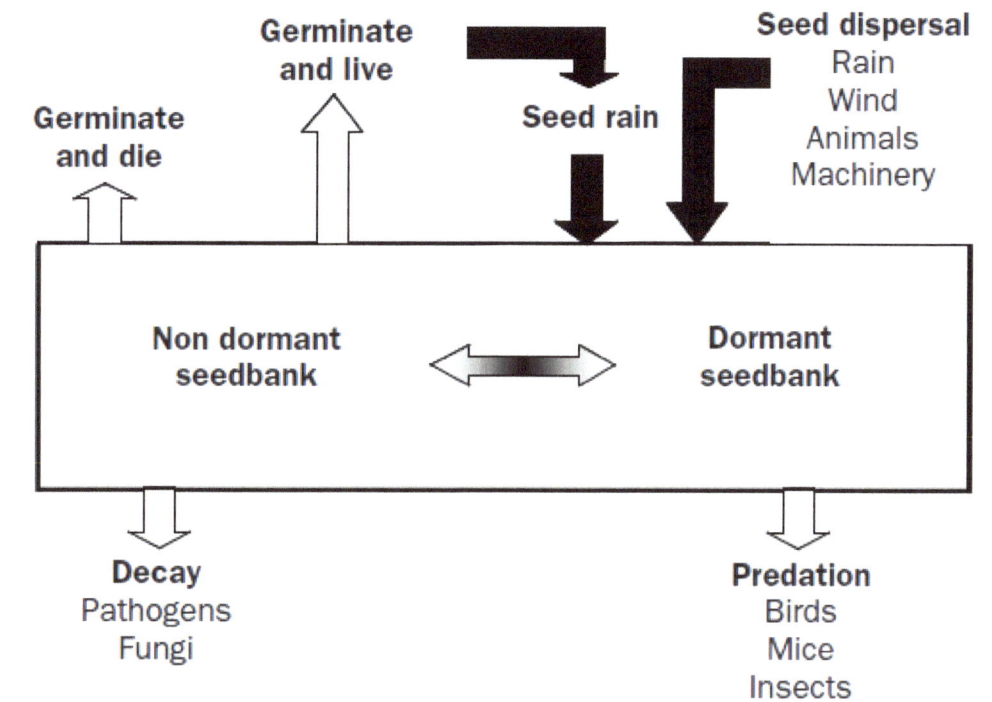

picture, you will come up with the answer. Water-impermeable seeds have either physical dormancy or a combination of physical and physiological dormancy, with the physical being the most common. Seeds with physical dormancy have a water gap in the seed coat that opens in response to an environmental signal, thereby allowing water to enter. When disturbance brings seeds to the soil surface, temperatures that are higher than those in the soil can cause the water gap to open. Consequently, the water gap indirectly serves as a depth sensor.

n case the seed decays, Nature does not plough and employs the earthworm, soil bacteria, together with deeply penetrating roots, to do the job. Nature does not supply water-soluble minerals to the soil; she ensures an automatic and ample application of organic matter which, in the process of decay, produces organic acids to act upon soil minerals and so make them capable of absorption by plant roots. Because we have failed to follow the example of nature we find that the soil in our care has apparently become incapable or 'deficient' of providing sufficient food to sustain our population in health. In modern farming, both crop production and livestock feeding, we are concerned with the provision of prepared nutrients imported to the farm, instead of making full use of the complete provisions of nature. The result is that we have burdened farming with the colossal cost of chemical fertilizers, sprays, insecticides, vaccines and medicines, while nature quietly continues to beat us, in the matter of both abundant production and healthy crops and animals, at no cost.

Problems of 'soil deficiencies' as far as the main elements are concerned, have only arisen with the increasing failure to acknowledge and act upon the natural laws. Without adequate decaying organic matter to release, in the process of its decay the otherwise non-available phosphates, potassium and nitrogen, that we transport from sources of concentrated supply and by treatment with chemicals, render them water soluble. In powder form these water-soluble elements are then applied, to upset the natural balance of the soil, to impregnate the water particles of the soil with concentrations far in excess of the optimum natural supply. Upon these the plant draws, instead of utilizing the more slowly available organic elements of the humus.

Phosphate deficiency is one of the outstanding fallacies of science (in soil as distinct from certain types of solid rock). A soil only becomes 'deficient' when there is insufficient decaying organic matter upon it to release the mineral nutrients already present in an unavailable form, and gather them from the air and falling rain. The solution, therefore, is the adequate organic matter in the right place [Turner, 2009].

ich in minerals (magnesium, copper, zinc, selenium, others) seeds also posses a battery of invaluable vitamins, antioxidants, phytosterols (which lower cholesterol), omega-6, and omega-3 fatty acids. Let's take the apricot kernels for the example. Apricot seeds are the best source of Laetrile, vitamin B-15, and B-17. Placed in the cold storage they retain freshness for over a year. The kernels also contain cyanide, but in modest doses it does not harm. Kernels should be cracked just before you use them, otherwise they lose their potency.

aetrile (amygdalin or Vitamin B17) therapy is one of

the most popular and best known alternative cancer treatments. Laetrile works by targeting and killing cancer cells leaving the healthy cells intact, and building the immune system to fend off future outbreaks of cancer. It uses two different methods for killing cancer cells: a strict diet (as do all cancer treatments) and several supplements.

In order to supply our diet with anti-cancer, anti-aging means, antioxidants, immuno-modulators, we daily take a number and vast varieties of seed oils available in each pharmacy (grape seed oil, flax seed oil, watermelon seed oil, pumpkin seed oil, or apricot seed oil), and that are rich in essential fatty acids, such as linoleic acid (LA), gamma linolenic acid (GLA, or omega 6), alpha linolenic acid (ALA, or omega 3), laetrile (amygdalin). But for making these supplements work efficiently, we have to take them with the following transporters crucial for their utilization: Vitamin C, A, B-group (including Pangamic acid-B15), zinc, selenium, among many others. Failure to take the necessary transporters is the guarantee of serious metabolic disturbances. We yet are unaware of the full list of necessary transporters. Moreover, we have to know and avoid risky and harmful combinations: for example, laetrile cannot be taken in conjunction with blood thinners (Vitamin K, Warfarin), proteolytic or pancreatic enzymes (Vitalzym), Prebiotics (Activia, L-acidophilus, B-bifidum), and Bob Beck Protocol electromedicine devices (the blood purifier and magnetic pulser). Nature suggests a simple and efficient alternative. If you want to nourish your body with essential oils, return the seeds to the earth. Supply the soil with the natural elixir, and the soil will supply you back a

healthy harvest with the balanced content and combinations – all in smart proportions. Go for it!

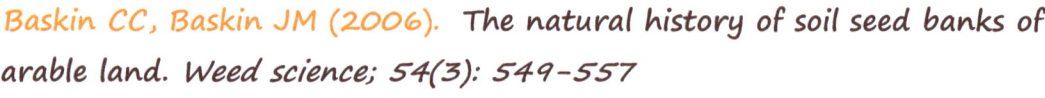

EFERENCES:

Baskin CC, Baskin JM (2006). The natural history of soil seed banks of arable land. *Weed science; 54(3): 549–557*

Beattie AJ, & Culver DC (1981). The guild of myrmecochores in the herbaceous flora of West Virginia forests. *Ecology, 62, 107–15*

Bullock JM, Kenward RE, Hails RS (2002). Dispersal ecology. Blackwell Science, Oxford UK

Giladi I (2006). Choosing benefits or partners: a review of the evidence for the evolution of myrmecochory. *Oikos, 112 (3), 481–92*

Gorb E, Gorb S (2003). Seed dispersal by ants in a deciduous forest ecosystem. *Springer Press (Paperback)*

Ibid, *p.16, p. 60.*

Lengyel S, Gove A, Latimer A, et al (2009). Ants sow the seeds of global diversification in flowering plants. *PLoS ONE, 4 (5)*

Menalled F (2008). Weed seedbank dynamics & integrated management of agricultural weeds. *Montana State University Guide.*

Moyle R., Filardi C, Smith C, et al (2009). Explosive Pleistocene diversification and hemispheric expansion of a "great speciator." *Proc. Nat. Acad. Sci. USA, 106 (6), 1863-8*

Plaseski T, Noveski P, Popeska Z, et al (2011). Association Study of Single Nucleotide Polymorphisms in FASLG, JMJDIA, LOC203413, TEX15, BRDT, OR2W3, INSR and TAS2R38 Genes with Male Infertility. *Journal of Andrology, October 20th*

Turner FN (2009). Fertility farming. *Journey to Forever, Faber & Faber Lmt, London [Paperback]*

Vog PH, Edelmann A, Kirsch S, et al (1996). Human Y chromosome azoospermia factors (AZF) mapped to different subregions in Yq11. *Human Molecular Genetics; 5(7): 933-943*

Wichmann MC, Alexander MJ, Soons MB, et al (2009). Human-mediated dispersal of seeds over long distances. *Proceedings of the Royal Society B; 276 (1656): 523-532*